Chem Trails are no Con (trail)

But are a con-troversy and con-spiracy
By Steve Holmes

"Con trail" is an abbreviation for "condensation trail". "Chem trail" is an abbreviation for "chemical trail." Con trails come from aircraft flying high in the sky because the aircraft engines create water vapor as a result from combustion of the fuel with the vapor condensing behind the aircraft but dissipating not far behind the aircraft on a day with relatively low humidity.

An aircraft with the following con trail is circled in this picture. As the craft moved to the left, the trail evaporated, so the trail never got very long.

On days with higher humidity, con trails can last longer.

Here is another aircraft with a con trail. As you can see, to the left of the cloud (behind the craft) the trail has dissipated but where there is the cloud, the trail is lasting longer and thus looks longer.

As shown here, most of these con trails have dissipated but where they went through a place where there is more humidity (as shown by the clouds), the trail lasted longer.

Some people don't believe in chem trails. They say that they are just con trails that happen to last longer. In my study of trails, it seems rather obvious to me that on a clear blue sky day, with no sign of higher humidity as shown by the slightest sign of any cloud, when the trail behind an aircraft never dissipates, crosses the whole sky, spreads out rather wide, and lasts for hours, it is not a regular con trail but a chem trail.

A problem is that it's hard to find an absolutely clear blue sky with chem trails (as shown above). So often, there are clouds somewhere in the sky and though the trails that frequently are parallel to one another and almost one atop the other, they can't be absolutely said to be chem trails (as shown below).

So, because of so big of a "gray area", common sense has to be used. In that I mean, if you see several chem trails almost atop one another, is that because that's a typical route for aircraft to fly? If there are two aircraft on close parallel paths making chem trails, is it normal for air traffic controllers to route aircraft so closely to one another? Here are some pictures to think about:

In the picture below, the trails aren't straight in some cases. Is it normal for aircraft (ones high enough to be long distance flights) to curve back and forth in the sky?

So, I told and even showed how some con trails last longer due to higher humidity. But is that an excuse for the trail to last for hours and keep expanding?

Here is a series of pictures of a con trail. The day that I took them, I spent the whole afternoon trying to spot any kind of aircraft trail in the sky but due to the many rain clouds in the sky, I couldn't see any trail until I finally saw this one later in the day. Since I know from previous observations that aircraft do cross the sky in the afternoon, if there were con trails that didn't dissipate, I'd have seen them, but there were none all afternoon. When I finally saw this aircraft with the con trail it was making (one that quickly dissipated behind the craft), it made it much more obvious that when I saw trails that hung around on cloudy days, they weren't con trails but chem trails.

As can be seen in the two pictures above, the trail is not long, dissipating a little behind the aircraft.

In the next three pictures of the series (on previous page), the con trail keeps dissipating a little behind the aircraft.

In the above picture, though the aircraft went by a cloud, the con trail didn't linger any longer than when the aircraft wasn't near the cloud.

In the above picture, again when the aircraft passed another cloud, the con trail did seem to linger a little longer as judged by the trail being a tiny bit longer.

In this above picture, the con trail seems to still be even a bit longer than in the last picture, maybe from more humidity there but again, it still dissipates not too far behind the aircraft.

In this last picture of the series (all taken December 2, 2014 in the Sacramento Valley of California), the con trail is still dissipating not too far behind the aircraft. So, to me this reaffirms that real con trails dissipate not far behind the aircraft, even when in the vicinity of clouds whereas when the trail grows so long as to cross the sky and lingers for hours, it is not a con trail and must be a chem trail!

There are a number of people on the Internet who want to "debunk" the theory of chem trails. They say that a long-lasting trail doesn't mean the trail isn't a con trail. Well, think about this and transfer what you personally have seen to the con/chem trail debate.

As these debunkers want it believed, aircraft trails that linger all day long are normal. So, since aircraft fly over my area quite frequently and I see both lingering and dissipating trails, ought they all be the same? Ought all of them be either lingering or dissipating, at least under the same circumstances (clouded sky or not)? Here are pictures taken December 4, 2014 (two days after the earlier series).

As these pictures above show, there are thin cirrus clouds, which ought to make con trails last for hours and cross the sky according to those who say all trails are con trail. The top left picture was taken at 5:59. As the top right picture (#2 of the series taken at 6:00) shows, the trail does get longer but where there aren't cirrus clouds the trail dissipates fast. Then in the middle left picture (taken 6:01) the aircraft is (shown in middle right picture taken 6:01) gone with the trail dissipating, but where there are cirrus clouds the trail lingers. A little later (still 6:01) the aircraft has a disappearing trail behind it but a gap. In the bottom right picture (take 6:02), the trail is still lingering where there are cirrus clouds, which shows how trails last longer when an aircraft goes through cirrus clouds but dissipate away from the clouds.

So, when you see a long-lasting trail that crosses the whole sky, lasting for hours, and there aren't cirrus clouds anywhere nearby, it is a chem trail.

**December 6, 2014 con trail when sky filled
with cirrus clouds**

If you breathe on a cold foggy day, you see your breath. That's a mini condensation cloud. If you walk down the street on that cold foggy day, do you see a trail of mini condensation clouds? No, because condensation "clouds" or trails dissipate.

Here is a screenshot of a YouTube talk about chem trails by a scientist:

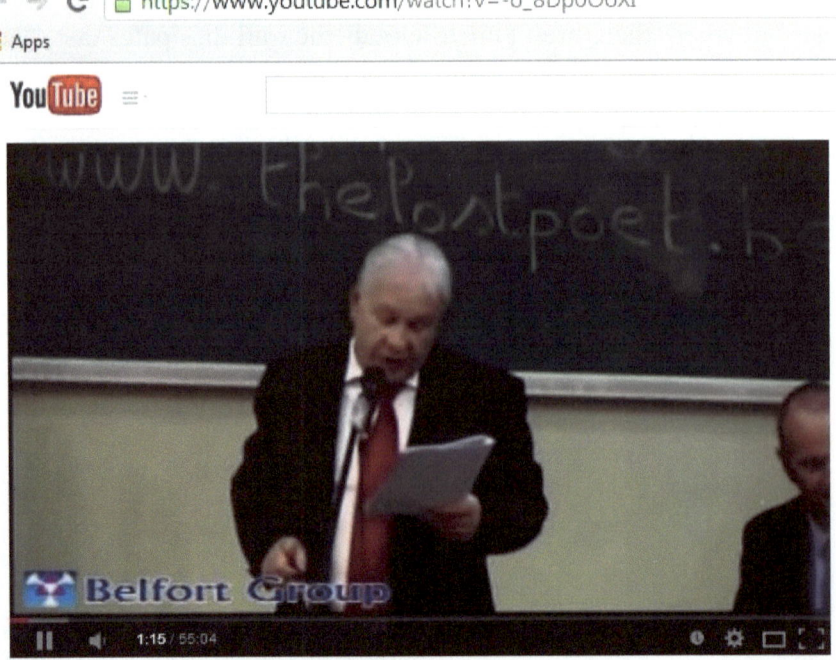

Aerospace Engineers as Whistleblowers Verify Global Geoengineering Operation with Chemtrails

Harold Saive

Why chem trails?

That is a big and a good question. Many people online have put forth their theories. There are two main theories. One theory is they're put over us to kill us with the chemicals that form the clouds. If it were just water vapor or condensation clouds, they would dissipate so to keep them in the sky longer somebody somewhere seems to have figured out a combination of chemicals to enable chem trails to stay in the sky.

Personally, this theory of the government trying to poison people doesn't sound right. Those trails are so high up and move with the winds so the aircraft would have to put the trail out over one area with the objective of having the chemicals "rain down" on another area to the east due to the winds moving the trails.

I think the likely reason is for weather modification. During the day, clouds cool an area. During the night, clouds keep earth's heat from radiating back into space, so that prevents cooling.

Many people say the earth is warming ("Global warming" became "climate change", which became "climate disruption".) To me, how the name of the "movement" changed from "global warming" seems to be an admission that the earth wasn't warming and that they had to change the name to be more accurate but still keep the people worried and on their side, willing to throw money wherever those in charge wanted the money to go. (NOTE: "weather" is a product on the Chicago mercantile exchange, so there is LOTS of money involved with how the weather is or isn't.)

Others point to data that shows the earth is cooling. The scientists have told us how "ice ages" come periodically and we're "overdue" for the next one. But that is a complicated subject in itself and this booklet is about chem trails and not exactly what their purpose is but whether they are created for whatever purpose and that such trails are not mere con trails.

What is in chem trails?

The late Dr. Ilya Perlingueri was first published on Global Research in May 2010 wrote an article about chem trails, saying this: "Over the past decade, independent testing of Chemtrails around the country has shown a dangerous, extremely poisonous brew that includes: barium, nano aluminum-coated fiberglass [known as CHAFF], radioactive thorium, cadmium, chromium, nickel, desiccated blood, mold spores, yellow fungal mycotoxins, ethylene dibromide, and polymer fibers."

Sample Type: Aqueous			
Sample Name:		Unlabelled Water Sample	
Lab Number:		788376.1	
Total Aluminium	g/m³	0.058	-
Total Arsenic	g/m³	< 0.0011	-
Total Barium	g/m³	0.0025	-
Total Boron	g/m³	0.0166	-
Analyst's Comments			

Australia Post Security tape applied and intact to lid of the container.

Rainwater in New Zealand, where chem trails were seen, was analyzed by a company specializing in analyzing water contents. Here is a graph of

their results: As the article on this pointed out, "barium should not be in rainwater at all."
https://clareswinney.wordpress.com/2010/05/09/chemtrail-chemicals-confirmed-barium-aluminium-found-in-whangarei-nz-rainwater/

I have read in other articles that strontium-90 is another chemical released in chem trails. Below is a YouTube video that talks of strontium also being found. In this video, it also tells how even Bill Gates is out saying that chem trails are needed to prevent Global Warming (do an internet search for "bill gates geoengineering" and you can find a 2012 article of how Gates wanted to use sulfate chemicals to accomplish that, despite that decades ago we were complaining about sulfates in smog). I don't believe in Global Warming and the name of the "movement" switched to "climate change" and then "climate disruption" so to me that tells pretty much that global warming was a fraud but, yes, I think they are manipulating the weather for another reason.

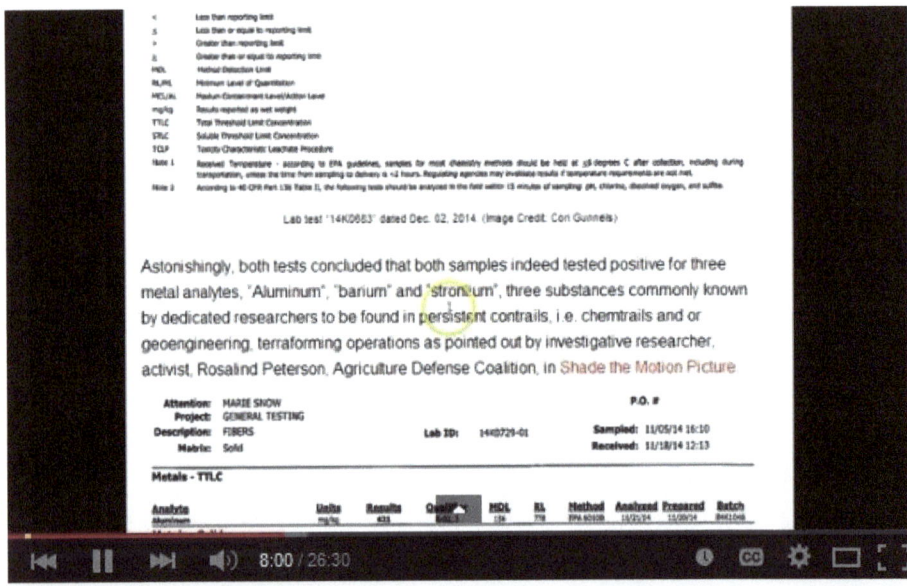

AZ Jet Drops Odd Fibers From Sky & the Stunning Test Results! Doc Proves DOD Can Experiment on Us

Here is a screen shot for a YouTube video that shows a report on an analysis of what was dropped over Arizona, which included aluminum, barium, and strontium

Conspiracy or not?

At dictionary.com the one definition of "conspiracy" is "an evil, unlawful, treacherous, or surreptitious plan formulated in secret by two or more persons; plot". So, are chem trails "evil, unlawful, treacherous or a surreptitious plan"? If they are to harm people, yes, they're "evil". I don't think they're "unlawful". Are they a "treacherous or surreptitious plan"? It's said that there are many HAARP (HAARP will be addressed later) stations around the earth, operated by several countries, and that HAARP works in conjunction with chem trails. I don't know if that is true but if it is and the countries are manipulating the weather, to stop or horrendously increase rain in another country, I think that would be considered "treacherous or surreptitious".

Is the "plan formulated in secret by two or more persons"? Definitely since there can't be just one person flying through the skies around the world putting out chem trails everywhere they're reported.

So, under that definition, yes, chem trails are a "conspiracy". Below is what Wikipedia says about the "conspiracy". To me, they just show their ignorance in not educating themselves about who is doing it (see section below).

Wikipedia has an article titled "Chemtrail conspiracy theory". It's not surprising that this mainstream publication wouldn't promote something that says the government is doing something nefarious. In their article they say, "Believers in the conspiracy theory speculate that the purpose of the claimed chemical release may be for solar radiation management, psychological manipulation, human population control, weather modification, or biological or chemical warfare, and that the trails are causing respiratory illnesses and other health problems." It goes on to say, "There are web sites dedicated to the conspiracy theory, and it is particularly favored by right-wing groups because it fits well with deep suspicion of government. In some accounts, the chemicals are described as barium and aluminum salts, polymer fibers, thorium, or silicon carbide." Here is a link to the article:
http://www.globalresearch.ca/chemtrails-the-consequences-of-toxic-metals-and-chemical-aerosols-on-human-health/19047

The book "Cloud Studies in Colour", 1967, has a picture of many chem trails side-by-side and calls them con trails. Here is a screen shot from a video that is trying to debunk chem trails:

Debunking "Contrails don't persist" with 70 years of books on clouds
___

The video then went on to show this from the book:

113. Condensation trails are left by aircraft when the air is sufficiently cold for the mixture of air and exhaust to be saturated (see 119). This does not usually happen except when the temperature is close to or below −40°C, in which case the cloud freezes almost instantaneously and does not readily evaporate. The cloud is then spread out by any wind shear which may be present. Below is a street of small cumulus.

What this shows me is that chem trails have been around since before 1967. The book or the person making the video that references the book doesn't happen to notice how many parallel trails there are in the same location and wonder how a particular route through the sky could be so popular in such a short period of time that the trails don't have much of any distance between them.

I wonder how the author of the above video would explain this other video:

ChemTrail Proof: 2 guys in Jet catches them spraying Chemicals Trails ~ Chem Trails ~ (mirror)

ChemTrail Proof: 2 guys in Jet catches them spraying Chemicals Trails ~ Chem Trails ~ (mirror)

The yellow ovals show jet's turbofans and the green ovals show the spray nozzles. Why is there anything coming out of nozzles? A little later the video shows how not only do the turbofans no longer have anything resembling any type of trail coming out of them but also the other nozzles:

Who is doing it?

From the evidence of government documents, the answer is that they are doing it. Here is a link to a 2005 document about weather modification:
https://www.govtrack.us/congress/bills/109/s517/text

Here is a screenshot of the site:

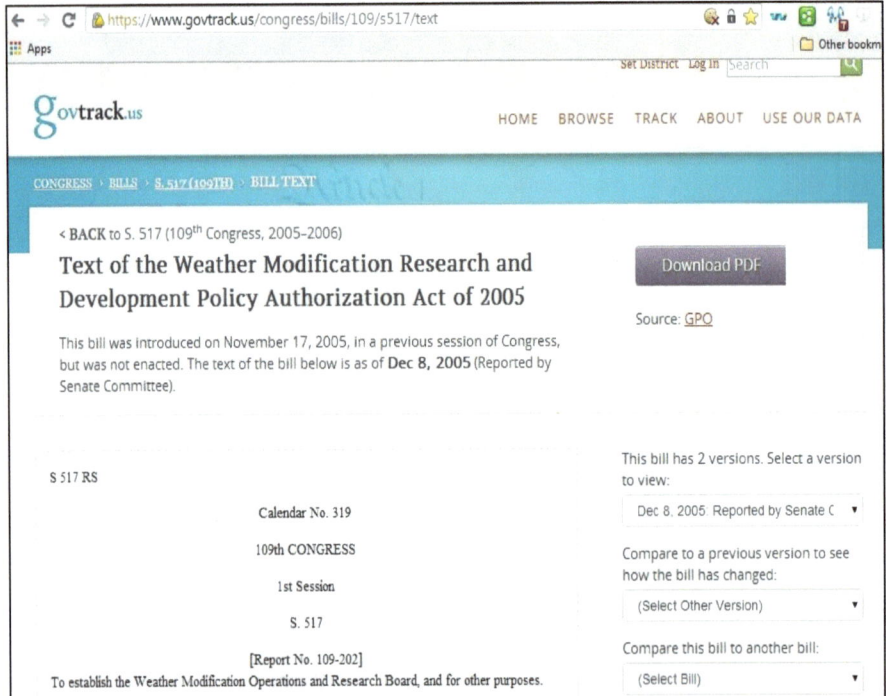

Right there in the title it says "Weather Modification Research" so if somebody says that the government isn't doing it and only those "wearing tinfoil hats" say such things, you ought to figure that source is not credible and doesn't want the truth to be known.

Here in "Sec. 3. Definitions." it tells how there will be "experimentation" and in that experimentation there will be "production and testing of models, devices, equipment, materials, and processes". Is there known to be any "testing" otherwise if chem trails aren't the manner of "experimenting"? Not that I have ever seen so this seems to be the "smoking gun" that the government IS "experimenting" with chem trails.

Here is another screenshot of the document:

(2) Coordination with relevant organizations that engage in weather modification research.

(3) Development through partnerships among Federal agencies, States, and academic institutions of new technologies and approaches for weather modification.

(4) Scholarships and educational opportunities that encourage an interdisciplinary approach to weather modification.

SEC. 3. DEFINITIONS.

In this Act:

(1) BOARD- The term `Board' means the Weather Modification Research Advisory Board.

(2) RESEARCH AND DEVELOPMENT- The term `research and development' means theoretical analysis, exploration, experimentation, and the extension of investigative findings and theories of a scientific or technical nature into practical application for experimental and demonstration purposes, including the experimental production and testing of models, devices, equipment, materials, and processes.

(3) tells of "new technologies", which means they aren't talking about just doing cloud seeding as they have done for decades. Here is a screenshot for part of what is in section 6:

(6) a description of the relationship between research conducted on weather modification and research conducted pursuant to the Global Change Research Act of 1990 (15 U.S.C. 2921 et seq.), as well as research on weather forecasting and prediction; and

To me "Global Change" looks awfully similar to "Global Warming" that so many want to blame on CO_2. If it's not CO_2 that they're talking about, what sort of changes are they? Here is a screenshot of what Wikipedia says about this act:

Yes, it's related to "research into Global Warming". My question is whether they are researching how to stop it or to make it happen (in case those who say earth is going into a cold cycle is true but withheld from the people). I note that nowhere in this Wikipedia article does it mention CO_2 being studied, so what is it all about?

While using an Internet search engine to find the above article, I also saw another link to a government article about the matter:

Global Change Research Act of 1990

From Wikipedia, the free encyclopedia

The template below (*Infobox U.S. legislation*) is being considered for merging. See templates for discussion to help reach a consensus.

The **Global Change Research Act** 1990 is a United States law requiring research into global warming and related issues. It requires a report to Congress every four years on the environmental, economic, health and safety consequences of climate change.

According to a summary by the Congressional Research Service, the Act:[1][2]

Global Change Research Act of 1990

← → C 🗋 www.law.cornell.edu/uscode/text/15/2931

⠿ Apps

15 U.S. Code § 2931 - Findings and purpose

Current through Pub. L. 113-185. (See Public Laws for the current Congress.)

| US Code | Notes |

prev | next

(a) Findings

The Congress makes the following findings:

(1) Industrial, agricultural, and other human activities, coupled with an expanding world population, are contributing to processes of global change that may significantly alter the Earth habitat within a few human generations.

(2) Such human-induced changes, in conjunction with natural fluctuations, may lead to significant global warming and thus alter world climate patterns and increase global sea levels. Over the next century, these consequences could adversely affect world agricultural and marine production, coastal habitability, biological diversity, human health, and global economic and social well-being.

(3) The release of chlorofluorocarbons and other stratospheric ozone-depleting substances is rapidly reducing the ability of the atmosphere to screen out harmful ultraviolet radiation, which could adversely affect human health and ecological systems.

(4) Development of effective policies to abate, mitigate, and cope with global change will rely on greatly improved scientific understanding of global environmental processes and on our ability to distinguish human-induced from natural global change.

I'm not sure of the exact date of this but it does reference "IMPROVING MEDICARE POST-ACUTE CARE TRANSFORMATION ACT OF 2014" so I presume it is an article from 2014 (the year of this booklet). What I note is that "chlorofluorocarbons and other stratospheric ozone-depleting substances" are mentioned but not CO_2 (but this is getting off-track from chem trails and the purpose of showing these government articles is to show that the government *is* involved in weather modification.

This document, below, outlines in great detail the existence of expanding US weather modification programs for at least a decade prior to the document in question. (Back to at least 1956). A "special commission" is outlined in this document to coordinate the multiple governmental agencies involved with US weather modification programs as well as independent contractors and universities which the report also mentions.

A less nefarious name for what is being done is called "geoengineering".

Here is an article that tells how, again, the US government is involved in it. This time it is the CIA: "CIA backs $630,000 study into how to control global weather through geoengineering" http://www.independent.co.uk/news/world/americas/cia-backs-630000-study-into-how-to-control-global-weather-through-geoengineering-8724501.html

Here is an article that tells how geoengineering is important to our military: "WEATHER AS A FORCE MULTIPLIER: OWNING THE WEATHER IN 2025 MILITARY APPLICATIONS OF WEATHER MODIFICATION" http://www.abovetopsecret.com/forum/thread59281/pg1

Here is a link to an Air Force chem trail document: http://saive.com/911/DOCS/Chemtrails_Chemistry_Manual_USAF_Academy_1990-OPT.pdf

Are other groups creating chem trails? Is the weather important to others? Of course, the weather is most important to the agriculture industry. Not enough water or too much water at the wrong time can be devastating on crops.

Here is an article that told how weather modification was used way back in 1952. "Rain-making link to killer floods" http://news.bbc.co.uk/2/hi/uk_news/1516880.stm

Here is another article: "Where Would You Like Your New Glacier?" http://www.ipsnews.net/2014/02/like-new-glacier/ Here is one sentence from the article: "The idea sounds like harebrained science-fiction, but the accelerated retreat of glaciers due to global warming and the effects of mining is leading scientists to seek to restore or recreate these valuable reservoirs of fresh water." It seems rather obvious to me that the restoring or recreating of glaciers would be involving climate change, which could involve chem trails.

CHEMTRAILS...
U.S. DEPT. OF
D 305.19:C 42 DEFENSE...
FALL 1990

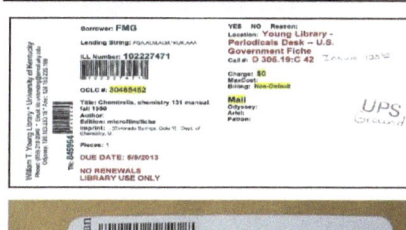

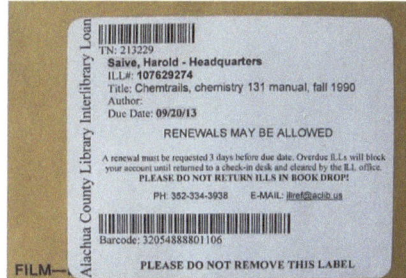

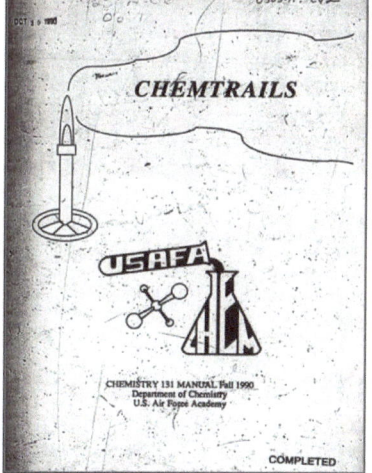

HAARP

"HAARP" is the acronym for "high frequency active auroral research program". Here is what Wikipedia says about it. "The **High Frequency Active Auroral Research Program** (**HAARP**) is an ionospheric research program jointly funded by the U.S. Air Force, the U.S. Navy, the University of Alaska, and the Defense Advanced Research Projects Agency (DARPA). Designed and built by BAE Advanced Technologies (BAEAT), its purpose is to analyze the ionosphere and investigate the potential for developing ionospheric enhancement technology for radio communications and surveillance.[2] The HAARP program operates a major sub-arctic facility, named the HAARP Research Station, on an Air Force–owned site near Gakona, Alaska."

The official HAARP web site says the purpose of HAARP is to do studies to help with communications that relate to the ionosphere.

Defense Hearing: Defense Research and Innovation (HAARP Segment)

Here is aYouTube video of a hearing where Dr. Walker said HAARP was used to "inject energy into the ionosphere to control it"

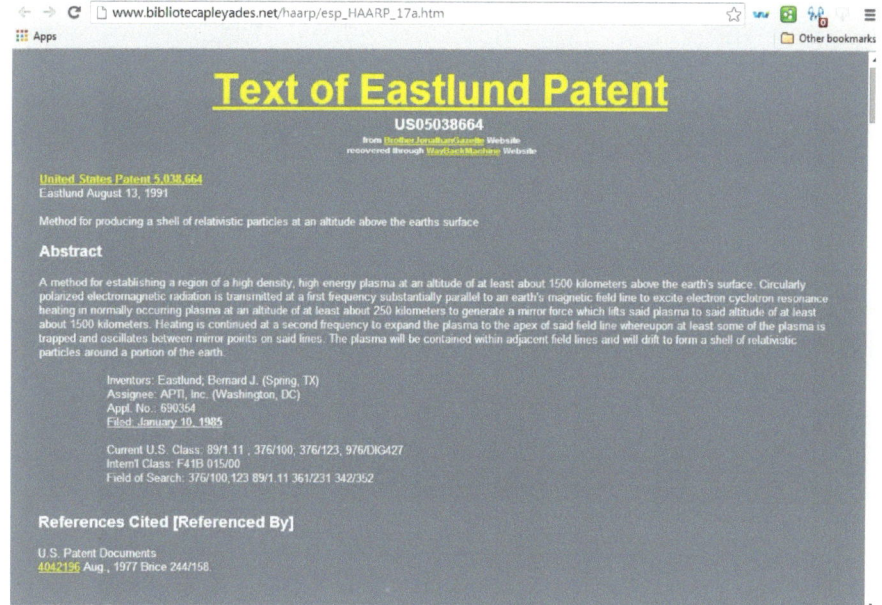

Here is a screenshot of the web page that tells of the patent for HAARP, the "Eastlund Patent"

It has been said by at least several people who "know" about HAARP that several countries have such a facility and are using it.
Alaska HAARP facility said by gov't to be shut down in June 2014.

This is the Alaska HAARP facility

Chemtrails Haarp (MUST WATCH)

In this YouTube video, a person complains about the chem trails from one horizon to the other, with hardly a real cloud in the sky, in March 2012

Here is a site that supposedly keeps track of HAARP over the USA: http://www.haarpstatu snetwork.com/ Here is their report for December 5, 2014:

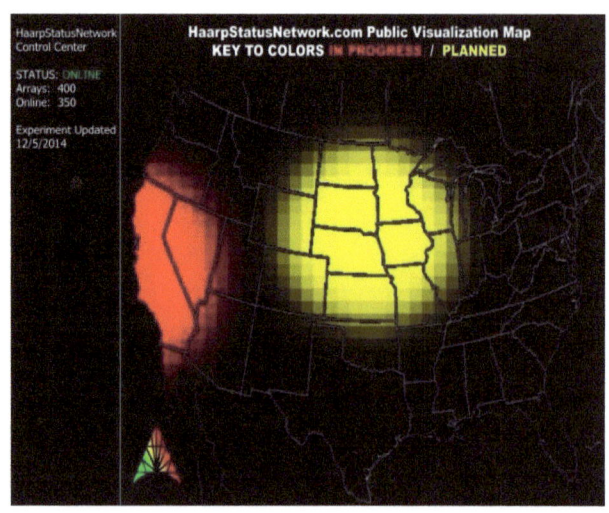

<u>Some other references:</u>

This online article has many references and links to documents: "Want to know about Chemtrails, HAARP, VLF, UHF and weather modification? Want to prove it to a non-believer? Here you go!"
https://sincedutch.wordpress.com/2011/10/04/want-to-know-about-haarp-vlf-uhf-and-weather-modification-want-to-prove-it-to-a-non-believer-here-you-go/

Another online article with lots of references links:
http://www.greatdreams.com/chems.htm

www.ingramcontent.com/pod-product-compliance
Lightning Source LLC
Chambersburg PA
CBHW041617180526
45159CB00002BC/891